Nelson In Mathematics
Workbook 1c

2nd edition

Name:

OXFORD
UNIVERSITY PRESS

To the student: Colour each circle as you complete the page to show how much work you have done in your book.

Content	Workbook pages
Number: count, count in twos, odd and even, compare, order	3 4 5 6 7
Measurement: time, hours, days, months	8 9 10
Number: place value, count in tens, multiples	11 12 13
Shape and space: direction, position	14 15 16 17
Number: add, addition strategies	18 19 20
Shape and space: 2D shapes, 3D shapes, sort shapes	21 22 23
Number: add, pairs that make 10, addition strategies	24 25 26
Measurement: money, coins, totals	27 28 29 30
Shape and space: half a shape, line symmetry	31 32
Measurement: time, time on the hour	33 34 35
Number: find half, share equally	36 37
Data: tables, lists, diagrams	38 39 40 41
Number: addition and subtraction facts, solve problems	42 43 44 45
Data: pictograms, block graphs	46 47 48

Revise numbers to 100

Date:

Write the number before and the number after.

Complete the flags.

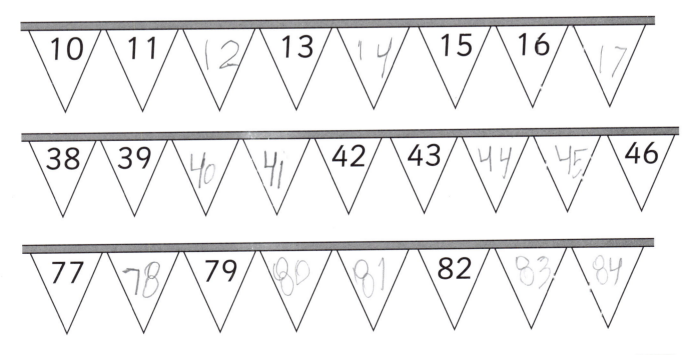

three 3

Counting on in twos

Date:

Start at 0. Count every other number.

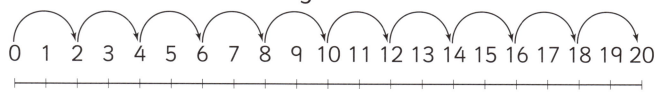

Start at 1. Count every other number.

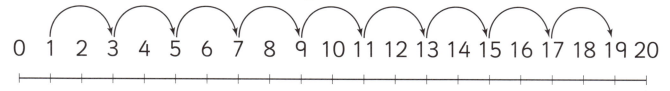

Count in twos. Write the numbers.

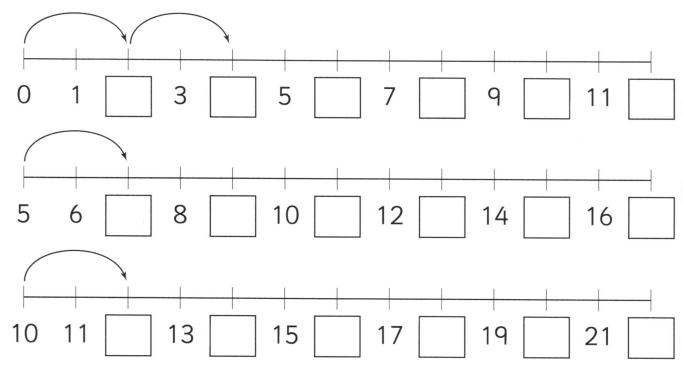

Write the number that is 2 more.

| 5 ___ | 9 ___ | 10 ___ | 11 ___ | 15 ___ | 18 ___ |

4 four

Count in twos

Date:

 1 pair = 2 shoes

  2 pairs = __4 shoes__

	3 pairs = __6__		4 pairs = __8__
	5 pairs = __10__		6 pairs = __12__
	7 pairs = __14__		8 pairs = __16__
	9 pairs = __18__		10 pairs = __20__

five 5

Two more, two less

Date:

10 + 2 = 12   10 − 2 = 8

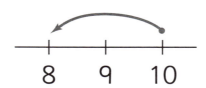

Write the numbers.

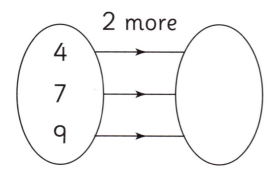
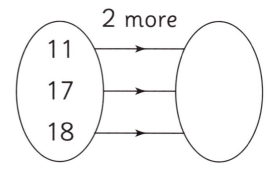
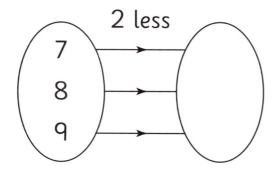
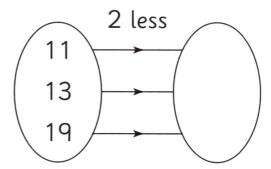

Nick has 8 sweets. Pete has 2 more.
How many does Pete have? _____

Maria has 9 stickers. Sonja has 2 less.
How many does Sonja have? _____

6 six

Even and odd

Date:

Even numbers can be shared into two equal groups.

Odd numbers always have one left over.

left

Write how many. Tick odd or even.

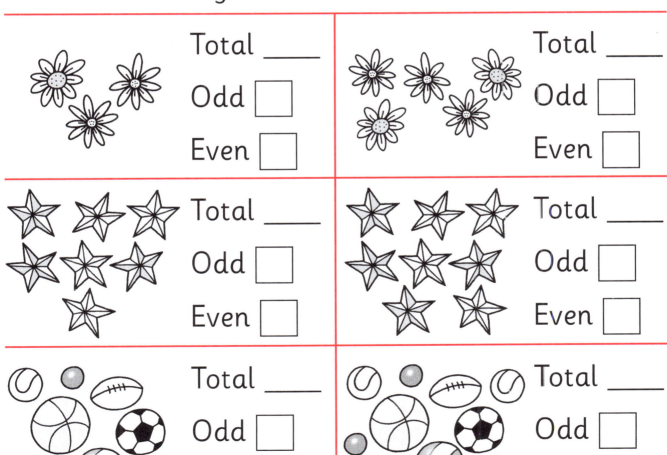

seven 7

Before and after

Date:

What do you do first?

Write before and after.

before   after

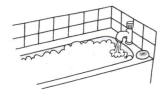     

_____ _____   _____ _____

_____ _____   _____ _____

_____ _____   _____ _____

8 eight

Hours

Date:

How much time do you spend on it?

less than 1 hour   about 1 hour   more than 1 hour

   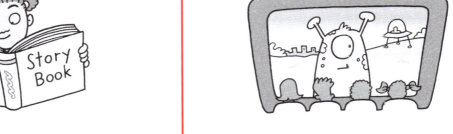

_____   _____

_____   _____

_____   _____

nine 9

Days and months

Date:

Order the days from 1st to 7th, starting with Sunday.

Order the months from 1st to 12th, starting with January.

Sunday
1st

Days

| Monday 1 | Wednesday 3 | Saturday 6 |

| Friday 5 | Sunday 7 | Tuesday 2 | Thursday 4 |

Months

| December | July | May |

| March | February | October |

| September | January | November |

| April | August | June |

10 ten

Tens and ones

Date:

Count the tens. Count the ones.
Write the number.

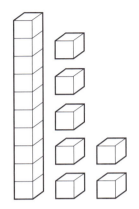

tens ☐ ones ☐

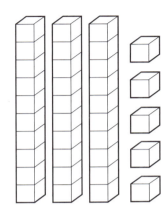

tens ☐ ones ☐

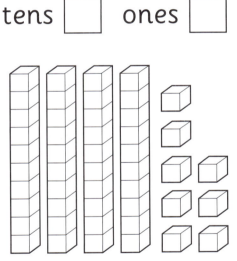

tens ☐ ones ☐

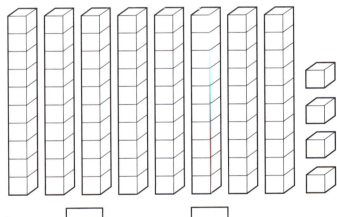

tens ☐ ones ☐

How many tens and ones?

23 = __ tens + __ ones   39 = __ tens + __ ones
55 = __ tens + __ ones   60 = __ tens + __ ones
74 = __ tens + __ ones   100 = __ tens + __ ones

eleven 11

Counting in tens

Date:

Count in tens to 100. Colour each number you count.

Say the numbers you counted.

Start at 4. Count in tens. Colour each number you count.

Say the numbers you counted.

1	2	3	4	5	6	7	8	9	10
11	12	13	14	15	16	17	18	19	20
21	22	23	24	25	26	27	28	29	30
31	32	33	34	35	36	37	38	39	40
41	42	43	44	45	46	47	48	49	50
51	52	53	54	55	56	57	58	59	60
61	62	63	64	65	66	67	68	69	70
71	72	73	74	75	76	77	78	79	80
81	82	83	84	85	86	87	88	89	90
91	92	93	94	95	96	97	98	99	100

Count in tens.
Write the missing numbers.

0   10   ___   ___   40   ___   60   ___

3   13   ___   33   ___   ___   63

Use the 1 to 100 chart.
Circle the number that is 10 more than:

12   25   44   61   90

Put a cross through the number that is 10 less than:

19   23   47   60   99

12 twelve

Ten more, ten less

Date:

Complete the table.

10 less		10 more
12 − 10 = 2	12	12 + 10 = 22

10 less		10 more

thirteen 13

Position

Date:

Where is the cat? Choose the correct words from the box.

<u>on top of</u> the blocks

| in front of | behind | next to | under | on top of |

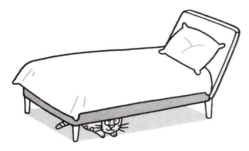

_____ the bed

_____ the TV

_____ the books

_____ the suitcase

_____ the car

_____ the bag

Movement and direction

Date:

Use the direction words.
Tell a friend how to get from the start to the finish.

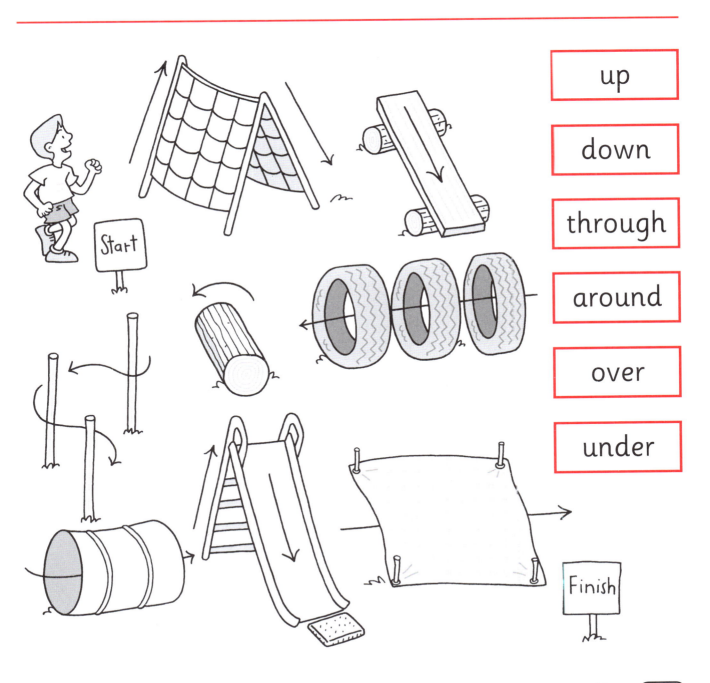

up

down

through

around

over

under

fifteen 15

Giving directions

Date:

Help the mouse to get the cheese.

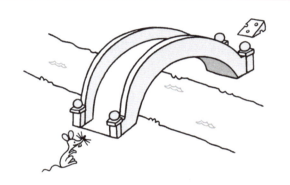

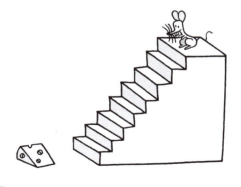

16 sixteen

Backwards and forwards   Date:

Must it go forwards or backwards to get into the correct position?

___backwards___

_____

_____

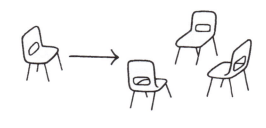

_____

_____

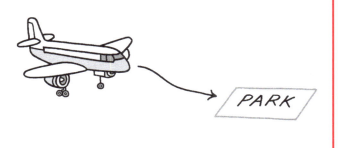

_____

_____

Faster adding

Date:

When you add, the order of the numbers does not matter. You can count on faster if you start with the bigger number.

$2 + 3$

$3 + 2$ $=$ 5

$4 + 5$
$5 + 4$ $=$

$2 + 7$
$7 + 2$ $=$

$6 + 1$
$1 + 6$ $=$

$3 + 6$
$6 + 3$ $=$

$9 + 1$
$1 + 9$ $=$

$4 + 1$
$1 + 4$ $=$

$9 + 3$
$3 + 9$ $=$

$2 + 5$
$5 + 2$ $=$

18 eighteen

Counting on to add

Date:

25 + 4 = __29__

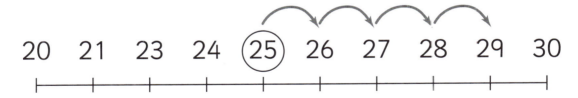

Use the number lines to help you find the answer to each sum.

0 1 2 3 4 5 6 7 8 9 10 11 12 13 14 15

16 17 18 19 20 21 22 23 24 25 26 27 28 29 30

10 + 5 = _____	17 + 3 = _____
19 + 5 = _____	23 + 5 = _____
26 + 4 = _____	12 + 2 = _____
16 + 3 = _____	18 + 8 = _____

More adding

Date:

Count on. Write the answers.

Add 4

9 → ☐
11 → ☐
14 → ☐
20 → ☐
24 → ☐

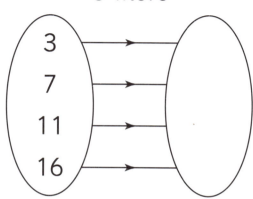

10
17
21
27
+3

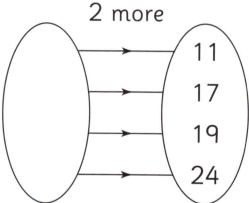

Add. Write the answers.

4 + 8 = ☐   2 + 9 = ☐

7 + 2 = ☐   3 + 8 = ☐

3 + 16 = ☐   7 + 13 = ☐

3 + 14 = ☐   2 + 17 = ☐

20 twenty

Naming and sorting 2D shapes   Date:

Draw lines to match.

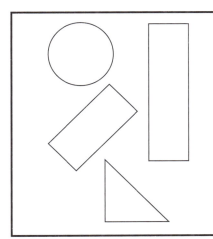

Circle

Triangle

Square

Rectangle

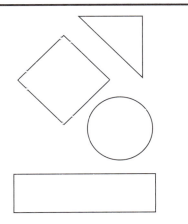

Colour the squares red.
Colour the rectangles yellow.

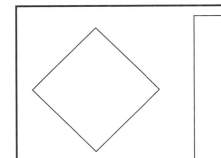

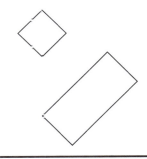

Draw a triangle inside a circle.	Draw a circle with a smaller circle inside it.

twenty-one **21**

Naming and sorting 3D shapes   Date:

Draw lines to match.

Cuboid	Cone	Cylinder	Cube

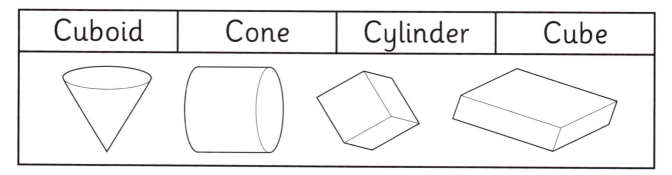

Colour the cubes red.
Colour the cuboids yellow.

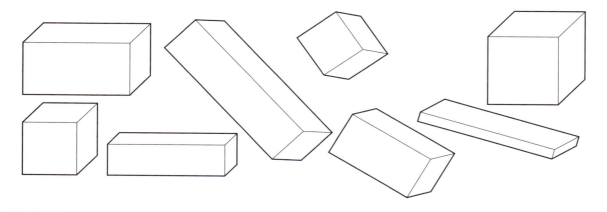

Colour all the shapes with only flat faces.

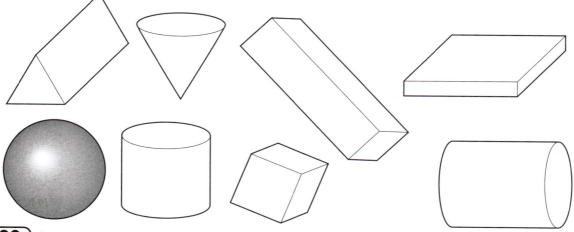

22 twenty-two

Sorting shapes

Date:

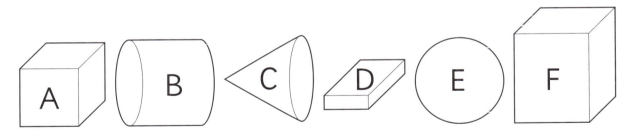

Write the letters in the correct blocks.

Shapes that have curved faces	Shapes that do not have curved faces

Shapes that have only flat faces	Shapes that do not have only flat faces

Shapes that have circular faces	Shapes that do not have circular faces

Shapes that have rectangular faces	Shapes that do not have rectangular faces

Shapes that have six faces	Shapes that do not have six faces

twenty-three

Adding more than two numbers   Date:

Add the blocks. Write the total.

   1 + 2 + 4 = ☐

   3 + 4 + 2 = ☐

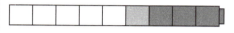   5 + 1 + 3 = ☐

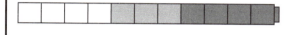   4 + 3 + 4 = ☐

   3 + 6 + 1 = ☐

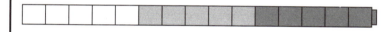   5 + 5 + 5 = ☐

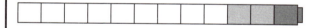   9 + 2 + 1 = ☐

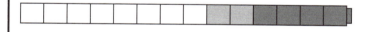   8 + 2 + 4 = ☐

   5 + 4 + 4 = ☐

   3 + 2 + 5 = ☐

   4 + 3 + 6 = ☐

   3 + 2 + 7 = ☐

Finding pairs that make 10   Date:

1 + 9 = 10   2 + 8 = 10   3 + 7 = 10
4 + 6 = 10   5 + 5 = 10

Look at this sum:
3 + 4 + 7

This is the same as: 3 + 7 + 4

3 + 7 = 10
10 + 4 = 14

So 3 + 4 + 7 = 14

Circle the two numbers that make 10 and write the answers.

(1) + 4 + (9) = 14   2 + 8 + 4 =

3 + 6 + 7 =   5 + 4 + 5 =

4 + 3 + 6 =   2 + 7 + 8 =

5 + 6 + 5 =   9 + 8 + 1 =

2 + 2 + 8 =   3 + 7 + 3 =

twenty-five 25

Adding in parts

Date:

8 + 3 = ☐

8 + 2 + 1 = 11

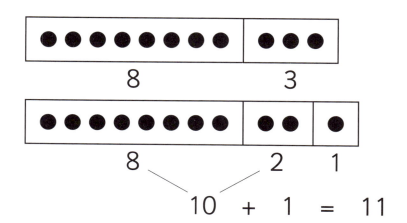

Write the missing numbers.

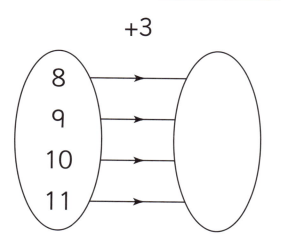
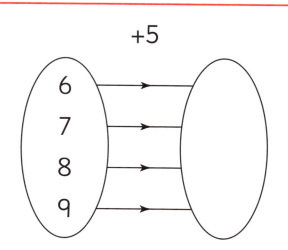
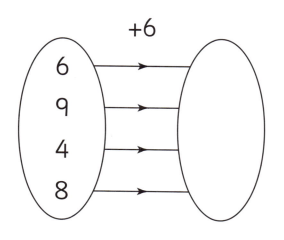
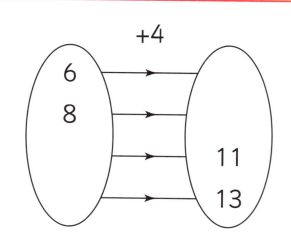

26 twenty-six

Coins in my country

Date:

Draw coins from your country in the table.

Value	Front of coin	Back of coin

twenty-seven 27

Making totals

Date:

Draw three coins in the first column.
Write the total in the second column.

Coins	Total value

How would you pay?

Date:

Draw the coins you could use to pay.

 5c

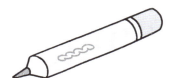 11c

 8c

 15c

twenty-nine 29

Make each amount

Date:

Make each amount in coins.
Draw two different ways.

 9c

 4c

 15c

 8c

30 thirty

Half of a shape

Date:

This cake is cut in half.
Each part is one-half.

half   half

The square is folded
to make two halves.

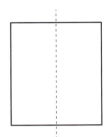 

half

Tick the ones that show half the shape.

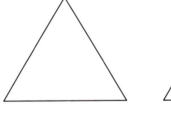    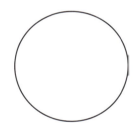 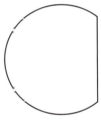

    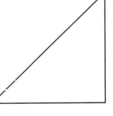

thirty-one 31

Symmetry

Date:

This is a line of symmetry.

This is not a line of symmetry.

Tick the lines of symmetry.

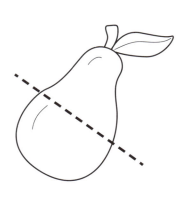

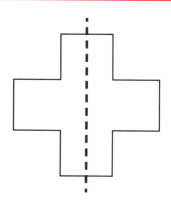
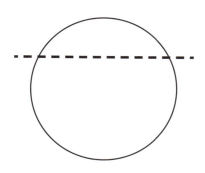
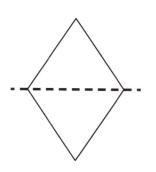

32   thirty-two

Write the time

Date:

Write the time.

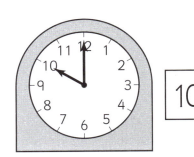

10 o'clock

 ☐ o'clock

 ☐ o'clock

 ☐ o'clock

 ☐ o'clock

 ☐ o'clock

 ☐ o'clock

 ☐ o'clock

 ☐ o'clock

thirty-three

Hands of the clock

Date:

Draw hands on the clocks to show the correct time.

7 o'clock

9 o'clock

2 o'clock

1 o'clock

10 o'clock

8 o'clock

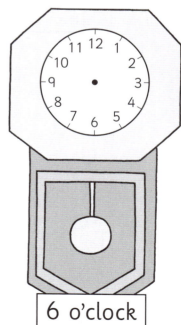

6 o'clock

34 thirty-four

Timetable

Date:

What time does it happen?
Complete the table.

| 7 o'clock | A |

7 o'clock	
8 o'clock	
1 o'clock	
4 o'clock	
6 o'clock	
3 o'clock	

thirty-five 35

Sharing

Date:

Circle half of each group.

Colour half of each group.

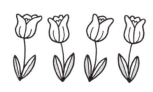

36 thirty-six

Share into groups

Date:

Share the flowers equally.
Draw the flowers in each vase.

thirty-seven 37

Sorting and classifying

Date:

Write how many there are of each item in the pet shop.

There are 5 rabbits.

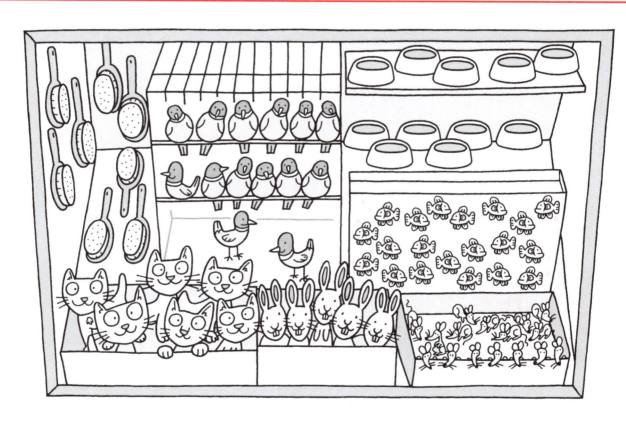

Sort in different ways

Date:

Sort the clothes in different ways. Explain how you sorted them.

Hot weather	A, B
Cold weather	
Any weather	

Hot weather	
Cold weather	
Any weather	

Girls' clothes	
Boys' clothes	
Either	

thirty-nine 39

How many cups?

Date:

You will need:

1 cup   1 big bottle   1 small bottle   1 bowl   1 pot   1 bucket

	What I used to fill it	How many?
cup	small bottle	less than 1
big bottle		
small bottle		
bowl		
pot		
bucket		

40 forty

Sorting diagrams

Date:

Write each number in the correct circles.

9   3   12   6   8   1   5   4   2   11

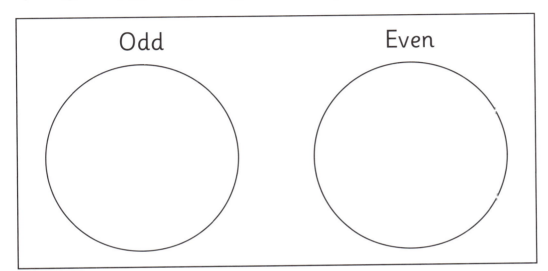

Write the letters of the shapes in the correct places.

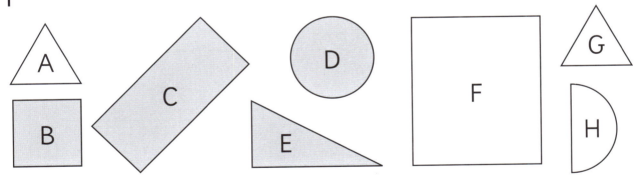

	Grey	Not grey
Rectangle		
Not rectangle		

forty-one 41

Number games

Date:

Throw the dice.
Use the dice numbers to make number sentences.

☐ + ☐ = _____

☐ − ☐ = _____

☐ + 5 − ☐ = _____

☐ − 1 + ☐ = _____

☐ + ☐ now double it = _____

☐ + ☐ now double it = _____

20 − ☐ + ☐ = _____

15 − ☐ + ☐ = _____

Addition and subtraction facts Date:

1 + 3 = 4   4 − 1 = 3
3 + 1 = 4   4 − 3 = 1

___3___ + ___2___ = ___
___ + ___ = ___
___ − ___ = ___
___ − ___ = ___

___2___ + ___4___ = ___
___ + ___ = ___
___ − ___ = ___
___ − ___ = ___

___ + ___ = ___
___ + ___ = ___
___ − ___ = ___
___ − ___ = ___

___ + ___ = ___
___ + ___ = ___
___ − ___ = ___
___ − ___ = ___

___ + ___ = ___
___ + ___ = ___
___ − ___ = ___
___ − ___ = ___

___ + ___ = ___
___ + ___ = ___
___ − ___ = ___
___ − ___ = ___

___ + ___ = ___
___ + ___ = ___
___ − ___ = ___
___ − ___ = ___

___ + ___ = ___
___ + ___ = ___
___ − ___ = ___
___ − ___ = ___

forty-three 43

Solve the problem

Date:

Write the number sentence and solve.
Use + or − and = .

 2 [+] 3 [=] 5

---

5 coconuts in a tree.
2 fall out.

___ [ ] ___ [ ] ___

---

6 birds on a branch.
3 more arrive.

___ [ ] ___ [ ] ___

---

4 people in a line.
1 more arrives.

___ [ ] ___ [ ] ___

---

8 worms.
Birds eat 3.

___ [ ] ___ [ ] ___

---

10 cherries.
4 get eaten.

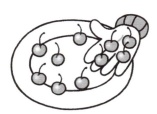

___ [ ] ___ [ ] ___

---

5 pencils.
7 children.
How many more pencils do they need?

___ [ ] ___ [ ] ___

44 forty-four

Test the rule

Date:

Rule	Test
You can add two numbers in any order and get the same answer.	3 + 5 = 8 5 + 3 = 8

Rule	Test
When you double a number, the answer will always end in 0, 2, 4, 6 or 8.	
If you draw pairs of dots to show an odd number, you will have one left over.	
When you add 1, you get the next whole number.	
When you subtract 1, you get the whole number that came before.	

forty-five

Fruit tally

Date:

Count each type of fruit.
Complete the graph.
Colour in 1 block for 1 fruit.

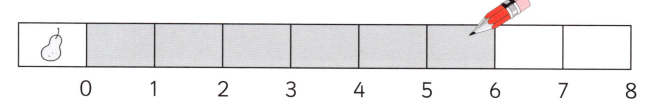

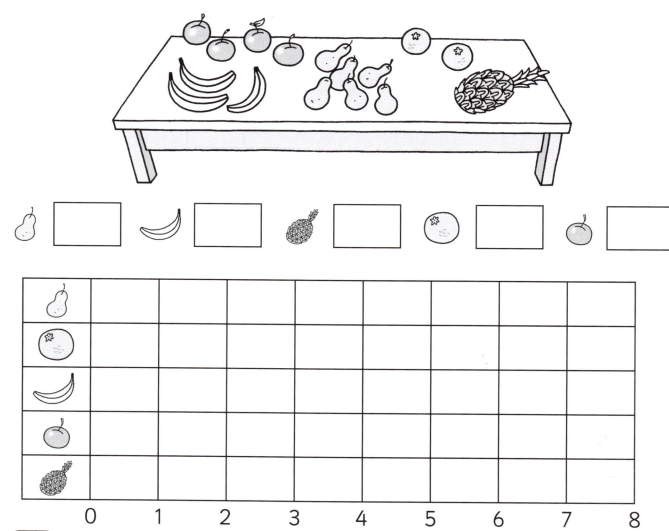

After-school activities

Date: _____

Read the graph.
Answer the questions.   ☐ = 1 student

After-school activities

| Sport | | | | | | | |
|ND |
TV							
Reading							
Dancing							
Games							

Which activity did most students do? __Sport__
How many students chose dancing? _____
Which two activities did equal numbers
of students choose? _____
How many more students chose reading
than dancing? _____
How many students chose TV and sport
altogether? _____

forty-seven 47

Draw a pictogram

Date:

Some children drew their favourite fruit.

Show this information on the graph.

Draw 👤 for one child.

Our favourite fruit

Apple 🍎					
Banana 🍌					
Orange 🍊					
Mango 🥭					

Key: 👤 = 1 child

48 forty-eight